HOW THEY LIVED

AN ICE AGE HUNTER

LUCILLA
WATSON

Illustrated by
John James

ROURKE ENTERPRISES, INC
Vero Beach, Florida 32964

First published in the
United States in 1987 by
Rourke Enterprises, Inc.
PO Box 3328, Vero Beach,
Florida 32964

First Published in 1986 by
Wayland (Publishers) Limited
61 Western Road, Hove
East Sussex BN3 1JD, England

Typseset by Kalligraphics Ltd, Redhill, Surrey
Printed in Italy by G. Canale & C.S.p.A., Turin

Library of Congress Cataloging-in-Publication Data

Watson, Lucilla
An Ice Age hunter.

(How they lived)
Bibliography: p.
Includes index.
Summary: Describes the day-to-day life of Ice
Age people and discusses their weapons, methods of
hunting, homes, clothing, and physical environment.
Includes a glossary of terms.
1. Paleolithic period—Juvenile literature.
2. Glacial epoch—Juvenile literature. [1. Man,
Prehistoric. 2. Glacial epoch] I. James, John,
1959– ill. II. Title. III. Series: How they
lived (Vero Beach, Fla.)
GN771.W36 1987 930.1′2 86–20270
ISBN 0–86592–143–1

CONTENTS

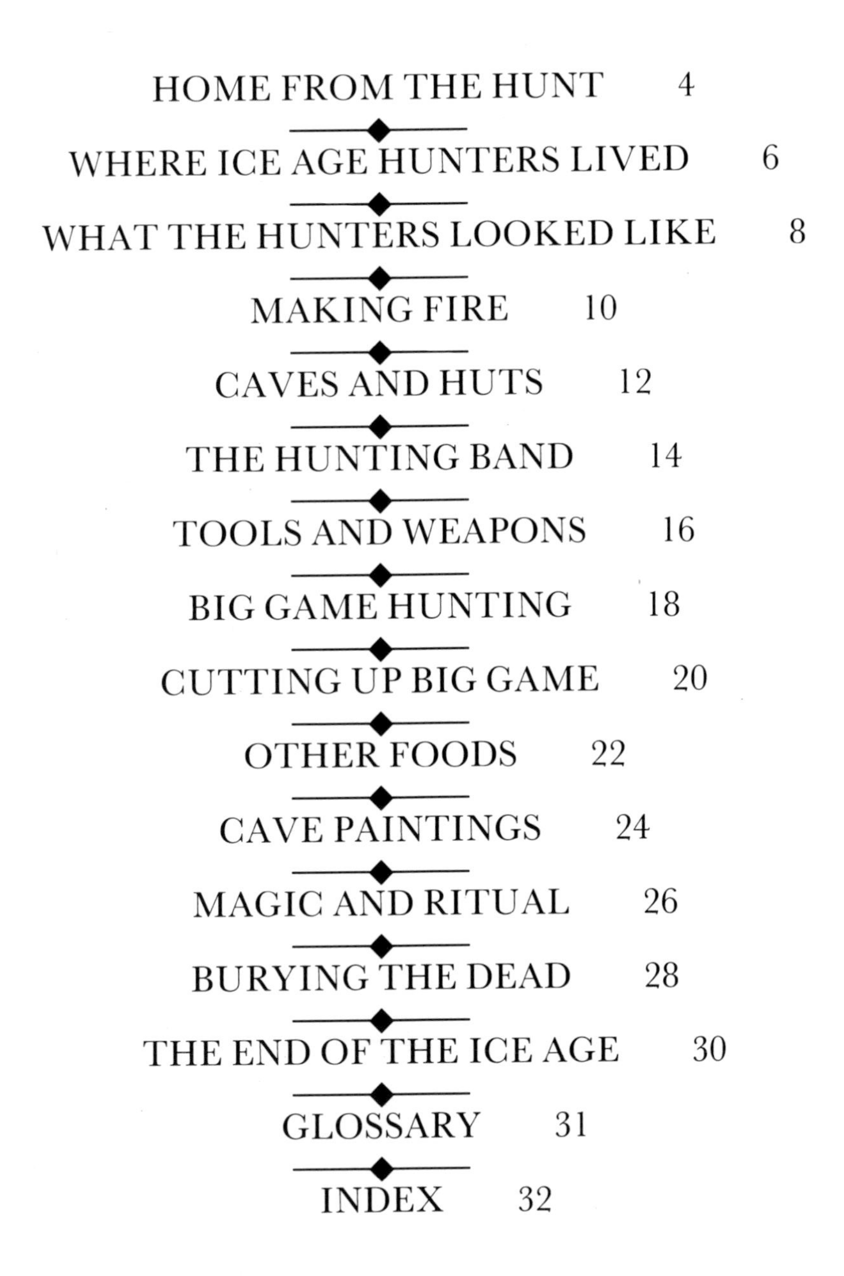

HOME FROM THE HUNT

The hunting band had been lucky. A mammoth had fallen straight into the trap that had been set for it. The men killed it quite quickly and now they were cutting it up into pieces to take home to cook.

One man was already carrying a piece of meat to the cave where his hungry family were waiting. He was cold and wet, the meat was heavy and it was going to be a long walk home, followed by a steep climb up to the cave on the hillside. But there would be a fire blazing, and the women and children would help cut the meat into smaller pieces and roast it.

The man was an Ice Age hunter. He lived 35,000 years ago, at a time known as the Ice Age, when countries like Britain, France, Spain and Germany were much, much colder and the sheets of ice at the North and South Pole were much bigger than they are today. No one really knows why the Ice Age started, nor why it stopped after 25,000 years. But we do know that it came and went very, very slowly, so that the people who lived at that time did not realize that it was getting colder and colder all the time, nor did they know that they were becoming Ice Age hunters.

Where Ice Age Hunters Lived

A map of Europe during the last Ice Age, showing the areas of ice, tundra and steppe, or grassland.

Ice Age hunters lived mostly in western and central Europe – what is now Spain, France and Germany, and right across to Siberia.

Because it was much colder in the Ice Age, the landscape looked different from that of today. It was too cold for trees like beech and oak to grow, so the landscape was quite empty. Only clusters of fir trees managed to grow here and there. There was no lush grass, but only stubbly plants and shrubs, and dry grass and moss. This kind of landscape is called tundra. There is still tundra in the northern parts of North America, Europe and Asia today.

The animals that lived on the tundra were also different from those we now find in western Europe.

During the Ice Age, you would have seen mammoths (hairy elephants with curly tusks), woolly rhinos, cave bear, bison, wolves, horses and herds of reindeer, just like the reindeer that live on the tundra of modern Greenland and Siberia.

Mammoths, woolly rhinos and cave bear are extinct now, but we know that these animals lived in Europe during the Ice Age because the Ice Age hunters painted beautiful pictures of them on the walls of their caves. The bones of these animals have also been found in caves, sometimes with cut marks in them and with the Ice Age hunters' stone knives nearby.

Above *A modern tundra landscape. Some animals of the Ice Age tundra are shown* **below.**

WHAT THE HUNTERS LOOKED LIKE

No one knows exactly what Ice Age hunters looked like. They did not paint pictures of themselves for us to look at. But they did leave their skulls and skeletons behind. Lots of these very old bones have been dug up and carefully examined by scientists, so we can make a good guess.

By examining the skulls, we can tell that an Ice Age hunter probably looked almost the same as we do. His face might have been more rugged,

From excavations like this one, scientists can guess what Ice Age people looked like.

and he certainly had bigger teeth than we have. He needed these for chewing large pieces of meat and grinding down tough roots and berries that he found to eat. His hair was also probably the same as ours, but he might have let it grow thicker and longer to help protect him from

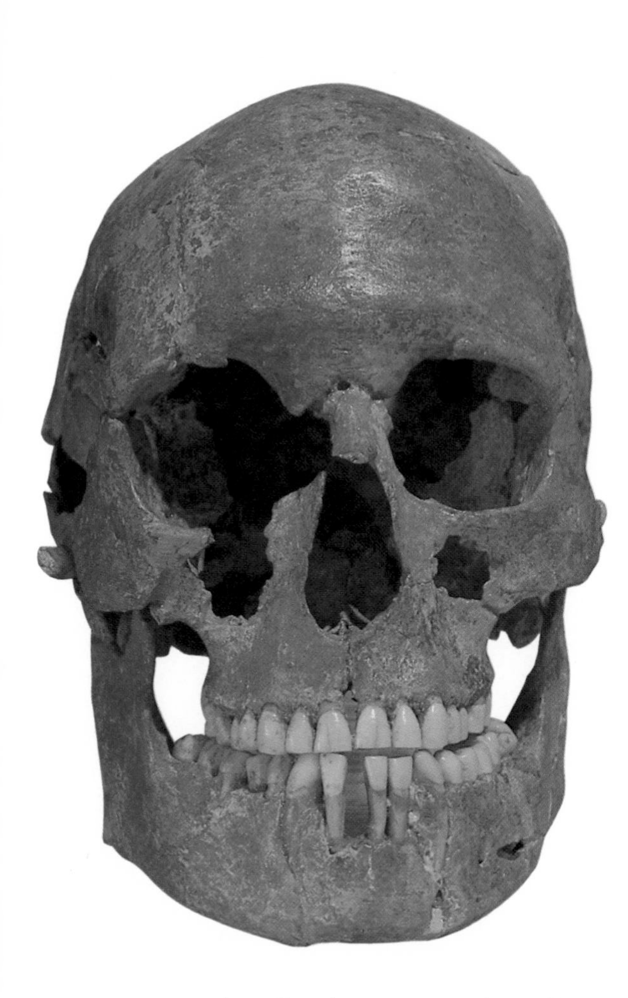

This skull of an Ice Age woman is quite similar to a modern one.

Ice Age hunters probably let their hair grow long to keep them warm.

the cold. He might also have had more hair than we have on his arms, legs and chest.

Ice Age hunters were also a little bit smaller than we are, and probably much stronger and hardier. They had to be to live in their cold climate with no houses or modern clothes. But Ice Age hunters were just as smart as we are, and we are going to look at some of the many ways they found to make their rugged, frozen life more comfortable.

Making Fire

Probably the most important thing in the life of the Ice Age hunters was fire. Without it, they could not have survived in the icy climate and their caves would have been dark, damp and frightening places.

Ice Age hunters used fire to keep themselves warm and dry, and probably for thawing, drying and cooking meat, and for melting snow and ice for water. A good fire would have scared off wolves, bears and other dangerous animals that would have stalked about in the darkness, or tried to come into the cave. Fire also lit the Ice Age hunters' torches when they explored the dark, twisting tunnels in the depths of their caves.

Nobody knows for sure how Ice Age hunters actually made fire. They probably gathered up some tinder of dry leaves and twigs. Then, they struck two stones together (a piece of flint and a piece of iron pyrite) to produce a spark and set fire to the tinder. They may have rubbed two dry sticks together until they smoldered.

Getting a fire going was obviously more difficult than just striking a match, like we do today. So once a fire was started, it must have been very important to keep it blazing. You can imagine the person with the job of keeping an eye on the fire, perhaps a young boy or old woman, getting into a lot of trouble with the rest of the family for letting it go out!

Fire, made by striking sparks from a piece of flint and iron pyrite, **above,** *was important for keeping warm and scaring off wild animals,* **opposite.**

CAVES AND HUTS

Caves were ideal places for Ice Age hunters to live. They were dry, a bit warmer than the air outside and they gave shelter from the weather. With a fire blazing at the entrance, caves must also have been welcome places to come home to after a day's hunting in the ice and snow.

An Ice Age hunter returns to his cave with a bird that he has caught for his family.

Most of the caves that Ice Age hunters chose to live in were either near a river, for fresh water, or up on a hillside, looking out on a valley. From a high vantage point the cave-dwellers could watch for herds of reindeer, or other animals, passing below.

In some caves, you can still see today the places where Ice Age hunters made their fires on the ground all those thousands of years ago. Their ancient hearths show up as blackened patches in the earth of the cave, sometimes with charred stones and bones around them.

Caves were probably rather untidy places, with animal bones and other waste – the remains of several hearty meals – scattered over the floor. But the warm corners of the cave where the hunters slept would have been swept clear of rubbish.

The hunters lived at the mouth of the cave. They only went down the dark, narrow corridors to paint pictures of animals on the walls and perhaps to perform strange rituals.

Not all Ice Age hunters lived in caves. In the summer, some of them might have followed herds of reindeer farther north, and camped in tents made out of skins as they went. Where there were no caves, Ice Age hunters might have lived in huts in winter as well as summer.

Right *Grotte Niaiux, an Ice Age hunter's cave in France, as it looks today.*

Below *When there were no caves available, people built huts.*

THE HUNTING BAND

We do not know exactly how many people lived in a cave. In the smaller caves, there might only have been room for a family of three or four, but the larger caves could have been the home of up to twenty-five people. They were probably all from four or five related families, and formed quite a large group, or band.

While the men of the band went out hunting, the women probably stayed at the cave with their children, or went out to pick berries and nuts and to dig for roots. Each woman might bear three or four children, but the likelihood is that only two of these children would have survived. When they grew up, the children probably married someone from a different band. Several bands might get together for special occasions, like a wedding or when someone died, but on the whole, different bands would not meet up very often.

Apart from the cold, the life of an Ice Age hunter was probably not very hard. It was less tiring than digging in the fields and much less boring than working in a factory.

On dark winter evenings, Ice Age hunters would have had plenty of time to sit around the fire making tools out of stone, carving bones into clever animal shapes and softening skins to make into clothes.

An Ice Age band at their camp. The women are scraping skins clean or making clothes, while the men prepare to go off hunting.

TOOLS AND WEAPONS

Ice Age hunters made many different kinds of tools out of flint, wood and bone. Each different kind of tool was made for a particular kind of job.

Probably the most important tools that Ice Age hunters made were spears for hunting big game. They made these spears by hafting, or fixing, a piece of pointed flint onto the end of a stick. They also used flint to make knives for cutting up the

Ice Age hunters were very clever with their hands, and could make fine tools and weapons.

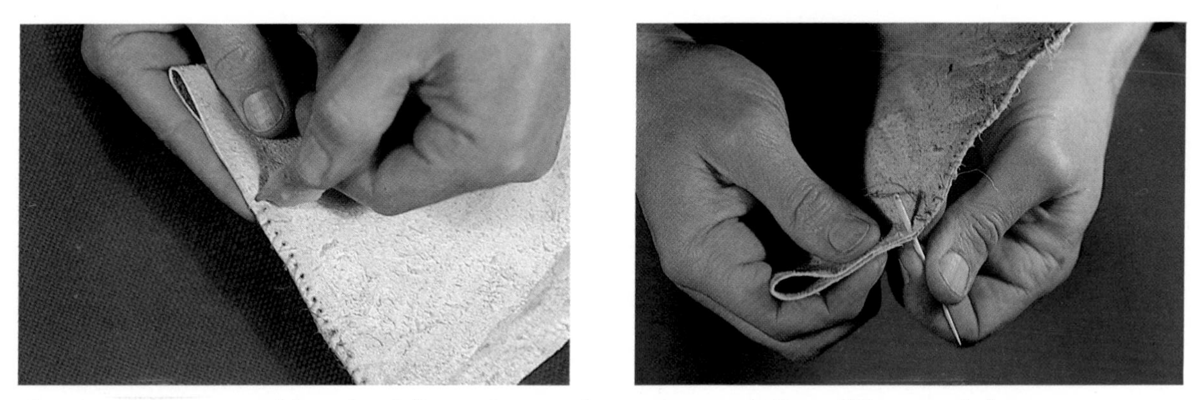

Ice Age hunters could make delicate borers from pieces of flint. They used these to make holes in the skins of the animals they killed, **above left.** *These were then sewn together using bone needles,* **above right.**

animals that they managed to kill, scrapers for cleaning animal skins before making them into clothes, and borers to make holes so that the skins could be sewn together.

The Ice Age hunter made these tools by knocking slender blades off a larger piece of flint with another piece of flint or with a hammer made out of reindeer antler.

They also liked to catch fish, and for this they made harpoons out of reindeer antler and fish hooks out of small pieces of bone. Reindeer antlers also made very good spear throwers, which the Ice Age hunters used to make their spears fly through the air faster and farther.

Sometimes they decorated their spear throwers with drawings of animals or made them into the shape of a horse or reindeer. They made these drawings by cutting lines into the antler with a small flint knife, called a burin. Ice Age hunters also used burins to decorate the walls of their caves with drawings of animals.

A spear thrower on which two bison fighting have been carved.

BIG GAME HUNTING

Of all the animals that roamed the icy, barren landscape 35,000 years ago, Ice Age hunters preferred to hunt the reindeer and the horse. Perhaps they were the easiest to catch. They were certainly not as dangerous as some Ice Age animals, like the enormous mammoth, with its tusks, and the woolly rhino, with its horn.

It was probably the younger men of the tribe who went out big game hunting, leaving the women, children and older people at the cave. One person on his own would have found it very difficult to catch a reindeer or a horse. So the men went hunting in groups. Before they set out, they must have had a rough idea of where to go to find herds of animals that they could kill.

Ice Age hunters probably hunted animals in several different ways. Once they had spotted a herd of, say,

reindeer, the hunters might have crept up on the herd, looking out for a young or weak animal to pick on. Then they would throw their spears at it until it collapsed. They may also have made traps for big game by digging a large, deep hole in the ground and covering it over with sticks and grass to hide it. The band of hunters would then drive a herd of, say, horses at a gallop into the trap where they could be killed easily.

Another way of getting close enough to large animals to kill them was to stampede them into natural bogs, where they would sink in and struggle helplessly. This was the best way of hunting very large animals, like mammoths and rhinos. The hunt must have been very exciting, but also quite dangerous. Sometimes hunters were killed or badly injured. You can imagine the rest of the band waiting anxiously at the cave for the hunters to come back, or watching the struggle from a safe distance.

A group of Ice Age hunters stalking a herd of reindeer. The hunters on the right are using spear throwers.

CUTTING UP BIG GAME

If the hunters killed a small animal, like a young reindeer or a foal, they might be able to carry it straight back to the cave. But if they managed to bring down a larger animal, like a horse, a reindeer or even a mammoth, they would have to cut up the dead animal and carry the pieces home separately.

Almost every part of the animal was used. Its meat was roasted over the fire. If it was a reindeer, its hooves were made into soup. Red-hot stones were dropped into leather bowls of water. Then the hooves were added and boiled. The bigger bones of the reindeer were made into spear throwers, the antlers into harpoons, and the smaller bones into awls and needles. Thin strips of leather could be cut from the skin to make thread. Buttons could be made from some of the bones, and even bone flutes have been found. The skin would also have been prized.

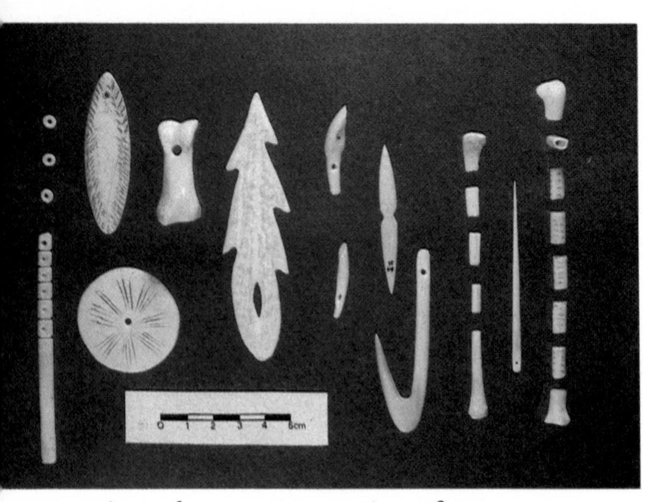

A modern reconstruction of some Ice Age bone jewelery.

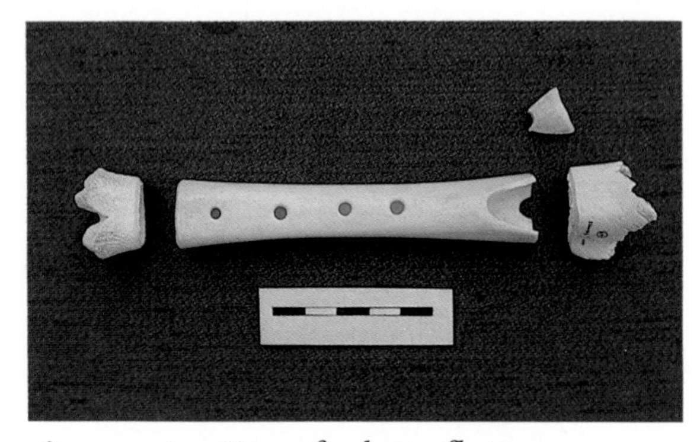

A reconstruction of a bone flute.

A large fur, like a cave bear's, would have made a warm rug for sleeping on or under. Lighter skins, like those of the reindeer, would have been made into clothes. But before the skin could be used, it had to be cleaned and softened with a flint scraper. Then holes were punched in it, with a borer, so that it could be sewn up to make clothes. Simple jackets were made for summer and anoraks and trousers for winter. Some Ice Age hunters may also have decorated their clothes with stone beads and little shells. When it was very cold, they might have worn boots and mittens, also made from the skins of animals they had killed.

Ice Age hunters used the skins of the animals they killed to make clothes.

OTHER FOODS

A good day's big game hunting meant a hearty meal for everyone in the cave that evening – and one more fur coat. If there was any meat left over, the hunters would probably keep it fresh and safe from scavengers by storing it under a mound of snow – the original frozen food.

But Ice Age hunters did not live on the meat of large animals all the time. Big game hunting may have been exciting but sometimes the animals would be difficult to catch, and the hunting band would return empty-handed.

Luckily, there were usually plenty of other things for them to eat. They speared fish with their harpoons, and may also have been able to catch birds and small animals. But when times were hard, they might have had to live on mice and insects.

During the summer and autumn, Ice Age hunters were able to go out and pick nuts, berries and other wild fruit to eat. They could also dig for tough but juicy roots to chew on. But they did not grow their own fruit and vegetables; they just went out into the woods and valleys to see what they could find.

It might have been the women and children of the band who had the job of gathering the roots and berries. Meanwhile, the men might have stayed behind in the cave, resting and getting ready for their next hunting expedition.

Below *An Ice Age harpoon (top) and spear point.*

Opposite *Women of the Ice Age gathering roots and herbs.*

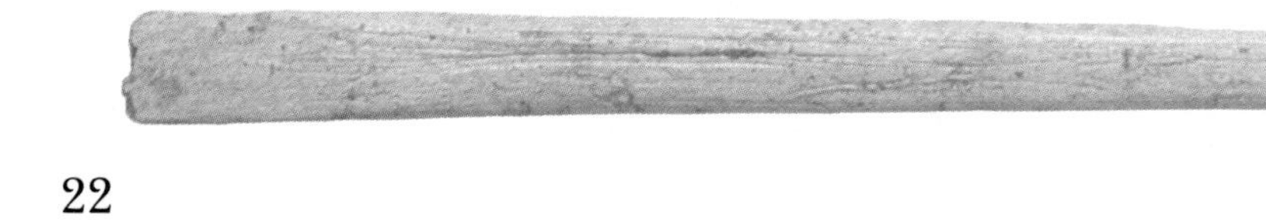

Cave Paintings

Ice Age hunters did not spend all their time hunting, cooking and making tools. They were also skilled artists who made beautiful paintings and drawings of animals on the walls of their caves. You can still see these paintings and drawings today because they have been preserved deep inside the caves. The most famous of these painted caves are to be found in France and Spain.

Mammoths, wolves, bison, bulls, horses and reindeer are just some of the animals that the Ice Age hunters drew and painted. They made paint by grinding colored rocks to a powder which they turned into a soft paste by mixing it with animal fat. The colors that they got in this way were brown, red, black, yellow and white. They painted either with their fingers or with brushes made from bristles stuck inside a hollow bone. Another way of getting the paint onto the cave wall was to fill a hollow bone with powder which was then blown onto the wall.

Some of the beautiful cave paintings at Lascaux, in France.

Red ochre was ground into a powder and used to make paint.

Instead of painting, Ice Age hunters sometimes made drawings by scratching, or engraving, with a flint burin. We know this because many blunt, used burins have been discovered inside the caves, just where the Ice Age artists left them – on the floor beneath the drawings on the cave wall.

Besides pictures of animals, the Ice Age hunters made other mysterious marks and scratches, which no one has managed to explain. They hardly ever made pictures of themselves – that is why we do not know exactly what they wore or what they looked like. But they often made silhouettes of their hands by placing them on the wall, blowing paint over them and then taking their hands away to leave clear outlines.

No one really knows why the Ice Age hunters painted their caves.

MAGIC AND RITUAL

The hunters probably performed rituals in the dark depths of their caves.

The Ice Age hunters did not decorate the mouths of their caves, where they spent their everyday lives. They made all their paintings and engravings deep inside the caves, where it was dark and mysterious. That is why many people think that the paintings were part of magic and ritual. Perhaps the hunters visited the paintings to bring themselves luck before setting out on a hunting trip and to help themselves feel stronger and braver when they came face to face with a fierce animal.

Some of the cave paintings are of men dressed as animals, wearing antlers on their heads or bison skins over their shoulders. These men – half human and half beast – may have been the shamans, or witch doctors. We can imagine them dancing curious ritual dances by torchlight in the depths of the caves.

Among the other mysterious objects left behind in the caves by the Ice Age hunters are little statues made of bone, stone or ivory. Some are of animals, but many are of rather fat women. The Ice Age hunters may have thought of them as goddesses because of their wish that the women would go on having children to keep the tribe alive.

An Ice Age cave sculpture of a woman with a horn.

BURYING THE DEAD

In the Ice Age, most people lived only about half as long as they do now. Almost half of all Ice Age children died before they had grown up. There may have been witch doctors and medicine men, who knew which herbs to give someone who was ill, but no one knew about germs or how to treat a wound or a broken leg.

When people died, they were carefully buried by the members of their hunting band. A grave was dug in the floor of the cave or beside the hut. The dead person was placed in

An Ice Age burial. The dead man's body is being covered with red ochre powder.

it, lying on his back or with his knees tied up under his chin. He was buried with all his clothes and necklaces on. The shaman, or perhaps everyone who was standing near the grave, sprinkled him with ochre, a red-colored rock ground into a powder. Then the body was covered with earth and the grave was filled in.

Sometimes, two or even three people were buried at the same time, and placed side by side in the grave. One Ice Age burial had as many as twenty bodies in it; perhaps some disease had killed almost everyone in one large hunting band.

An Ice Age grave site. The skull has been painted.

THE END OF THE ICE AGE

About 10,000 years ago, the Ice Age came to an end. The weather got warmer, and the sparse trees and grasses that covered much of western and central Europe gave way to thick forests. The kinds of animals changed too. The mammoths and woolly rhinos became extinct. The reindeer moved northwards to what is now Norway, Sweden and Finland. In their place, animals like deer, wild cattle and wild pigs came to live in the forests.

At about the same time, people stopped living in caves and gradually took up a whole new way of life. They started to live in houses. Instead of hunting big game and gathering fruits and wild oats, they began to tame wild pigs and cows and grow crops. So the world of the Ice Age hunter disappeared.

At the end of the Ice Age people gave up living in caves and built huts and primitive houses to live in. They also began to tame animals.

GLOSSARY

Antlers The horns of a deer.

Awl A small tool used for piercing holes.

Burin A small flint tool for drawing on rocks and bones.

Charred Blackened by fire.

Extinct Something that has died out.

Flint A kind of hard, sharp stone.

Harpoon A short spear with hooks down its point.

Hearth The place where a fire is made.

Ice Age The period of time 35,000 to 10,000 years ago, when it was much colder than it is now.

Iron pyrite A stone containing iron and sulphur.

Ritual A kind of prayer, sometimes with dressing up, singing and dancing.

Spear thrower A length of wood, bone or antler made to hold a spear and used to throw the spear farther.

Shaman A priest or witch doctor.

Stampede To frighten animals into a gallop.

INDEX

Picture acknowledgments

The pictures in this book were supplied by the following: Bryan and Cherry Alexander 7; British Museum of Natural History 22; Le Musée National de Préhistoire des Eyzies 9 (left), M.H. Newcomer 8, 17 (top right and left), 20 (both), 25; E.S. Ross 13, 24, 27, 29; Ronald Sheridan 17 (bottom).